# POSITIVE NUMBERS ZERO & NEGATIVE NUMBERS

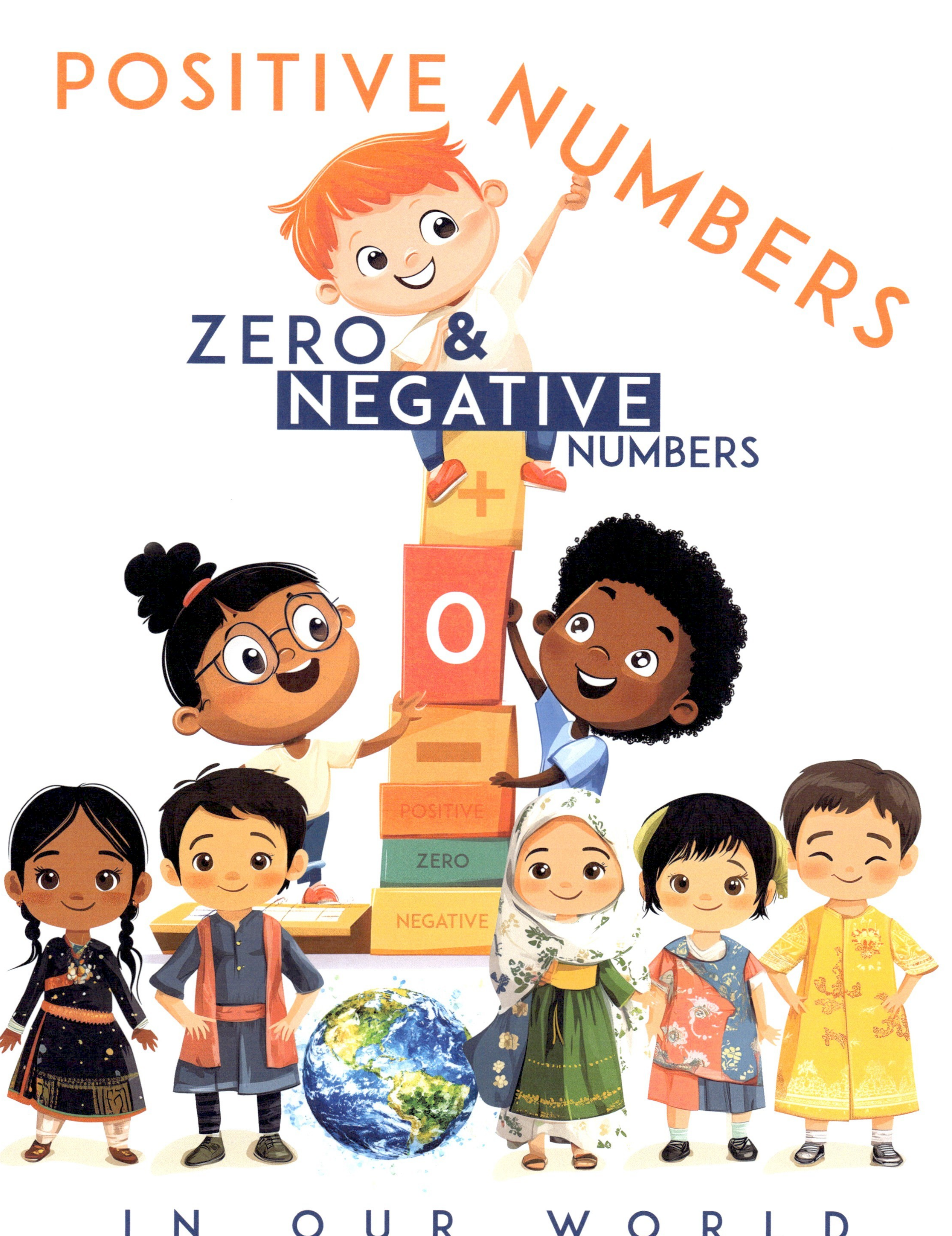


## IN OUR WORLD

PATRICIA LOBPRIES MCCARTY

Archway Publishing books may be ordered through booksellers or by contacting:

Archway Publishing
1663 Liberty Drive
Bloomington, IN 47403
www.archwaypublishing.com
844-669-3957



ISBN: 978-1-6657-6378-3 (sc)
ISBN: 978-1-6657-6379-0 (e)

Library of Congress Control Number: 2024915733

Print information available on the last page.

Archway Publishing rev. date: 01/30/2025

"Positive Numbers, Zero and Negative Numbers in Our World" introduces your child to a new description about numbers that they will soon be using in school mathematics. There are 20 scenarios which are beautifully illustrated by pictures, explained below the pictures and questions are asked pertaining to each situation.

The book introduces you to ideas illustrating numbers using a trip by an airplane, the height of the Empire State Building, body temperatures, gasoline in a vehicle, seeds in a garden, elevators in a building, speed limits near a school zone, the Lincoln Memorial steps, running speeds of animals, time travel and many more examples. Answers to some questions may change depending on the age and gender of the person reading the book!

# POSITIVE NUMBERS, ZERO AND NEGATIVE NUMBERS IN OUR WORLD

## WHAT IS A POSITIVE NUMBER?

A POSITIVE NUMBER IS ANY NUMBER THAT REPRESENTS MORE THAN ZERO OF ANYTHING.

POSITIVE NUMBERS ARE ON THE RIGHT OF ZERO ON A NUMBER LINE.

FOR EXAMPLE, RECEIVING $5 FOR YOUR BIRTHDAY OR GROWING 2 INCHES WOULD BE POSITIVE NUMBERS.

WORDS SYMBOLIZING POSITIVE NUMBERS ARE GAIN, RECEIVE, TEMPERATURE ABOVE ZERO AND ABOVE GROUND.

# WHAT NUMBER DOES ZERO REPRESENT?

ZERO IS A NUMBER REPRESENTING NO QUANTITY.

ZERO APPLES MEANS THERE ARE NO APPLES.

WORDS SYMBOLIZING ZERO WOULD BE GROUND LEVEL, SEA LEVEL OR NO MONEY IN BANK ACCOUNT.

# WHAT IS A NEGATIVE NUMBER?

A NEGATIVE NUMBER IS ANY NUMBER THAT REPRESENTS LESS THAN ZERO OF ANYTHING.

GIVING A FRIEND $5 FOR A BIRTHDAY PRESENT OR CATCHING A FISH THAT IS 2 INCHES SMALLER THAN ANOTHER FISH, WOULD BE NEGATIVE NUMBERS.

NEGATIVE NUMBERS ARE ON THE LEFT OF ZERO ON A NUMBER LINE.

NEGATIVE NUMBERS INDICATE A LOSS, DECREASE, GIVING, TEMPERATURE BELOW ZERO OR GOING DOWN IN AN ELEVATOR.

ON A NUMBER LINE,
POSITIVE NUMBERS APPEAR ON THE RIGHT SIDE OF ZERO AND NEGATIVE.

NUMBERS APPEAR ON THE LEFT SIDE OF ZERO.

<__-3____-2____-1_____0_____+1_____+2_____+3_______>

IF THERE IS NO SYMBOL IN FRONT OF THE NUMBER, THEN THE NUMBER IS POSITIVE.
FOR EXAMPLE, +2 = 2. NEGATIVE NUMBERS ALWAYS HAVE A SIGN, -3 = -3.

## ABOUT THE AUTHOR

PATRICIA LOBPRIES MCCARTY TAUGHT MATHEMATICS IN PUBLIC SCHOOLS, COLLEGES AND GIFTED AND TALENTED COURSES AT UNIVERSITIES AND COLLEGES. SHE CONTINUES TO TEACH MATHEMATICS AT A JUNIOR COLLEGE.

HER CHILDREN AND GRANDCHILDREN INSPIRED HER TO WRITE THIS BOOK. PATRICIA IS A NATIVE HOUSTONIAN WITH A DOG AND A CAT.

## ABOUT THE ARTIST AND DESIGNER

SLADE DELIBERTO: A MULTIFACETED INDIVIDUAL WHOSE LIFE JOURNEY EMBODIES RESILIENCE, CREATIVITY, AND A RELENTLESS PURSUIT OF EXCELLENCE.

FROM SERVING IN THE UNITED STATES MARINE CORPS TO BECOMING AN AWARD WINNING 3D ANIMATOR. HE HAS CREATED SOME OF THE MOST CUTTING-EDGE CONTENT FOR EXHIBITS. HE HAS CREATED ANIMATED AND EDUCATIONAL CONTENT, FOR ALMOST EVERY EXHIBIT AT THE HOUSTON MUSEUM OF NATURAL SCIENCE.

HIS PORTFOLIO DISPLAYS WORKING WITH LEADING TECH COMPANIES IN OIL AND GAS, RENEWABLES, ARCHITECTURE, SCIENCE, AI, VR, AUTOMOTIVE, ENTERTAINMENT, MUSIC, TV COMMERCIALS AND VISUAL EFFECTS FOR FILM.

# POSITIVE NUMBERS

The names of the fastest fish and the slowest fish both begin with a 'S'.

What are their names?
Sailfish and seahorse

## Zero

Zero miles per hour means the fish is not swimming.

## Negative Number

The dwarf seahorse is the slowest fish in the world with a speed of 5 feet per hour or 0.01 mph.

### Questions

**1. Why is the Indo-Pacific Sailfish so fast?**

The sailfish fold their fins back completely, and their bodies resemble a torpedo as they dash toward their targets. The largest sailfish are usually females.

**2. Why is the seahorse so slow?**

The seahorse is slow because of its rigid body structure, and it only has a small fin in the middle of its back to propel itself.

The dwarf seahorse uses its dorsal fin to create ripples that move it forward in an upright position, while its pectoral fins help it to steer.

Because of it's slow swimming speed, the dwarf seahorse cannot chase prey, so it instead waits for food like shrimp larvae to drift by.

## Fun Fact

The Indo-Pacific Sailfish reaches speeds of over 68 miles per hour.

It is the fastest fish in the ocean.
Sixty-eight mph equals + 68 mph = 68 mph.

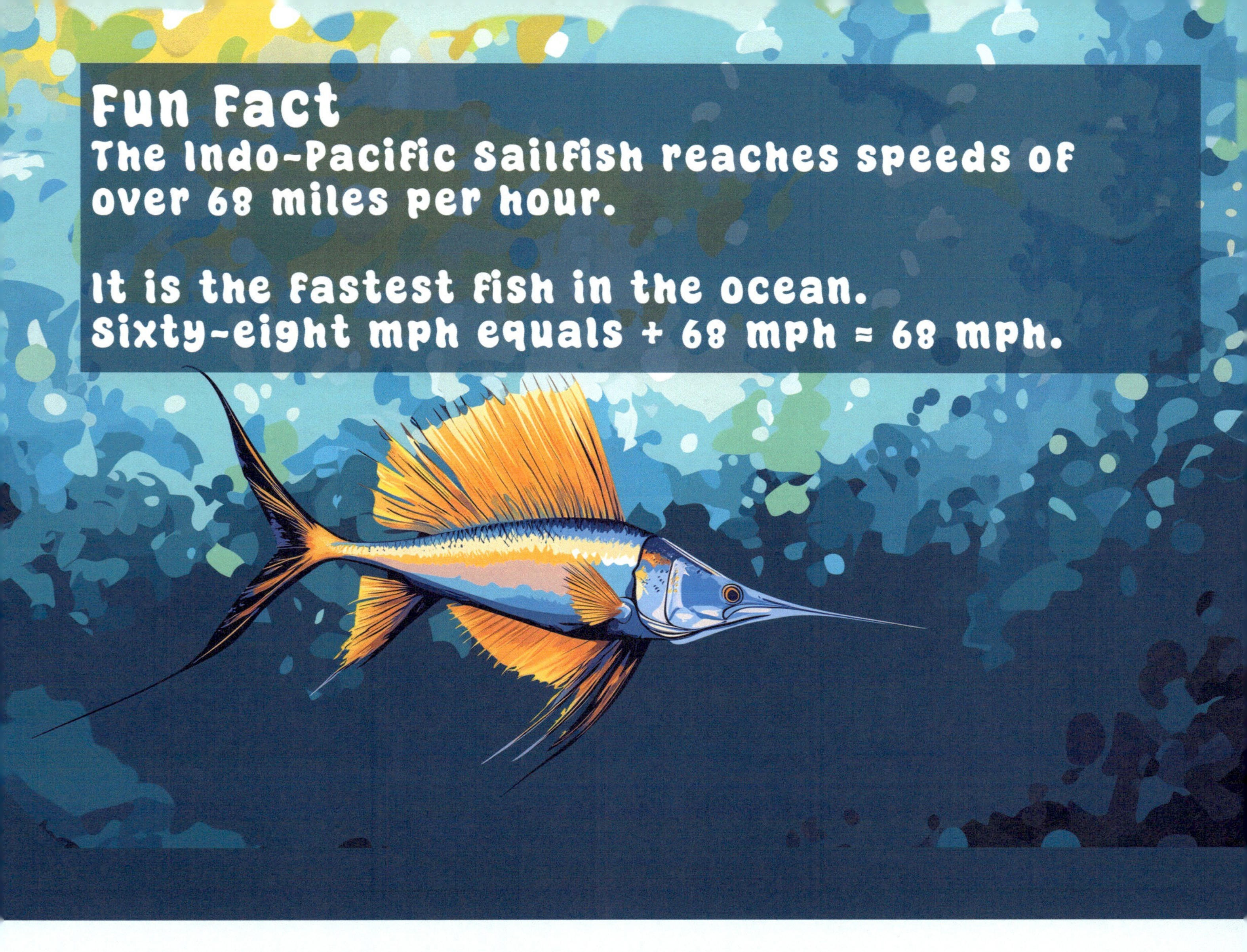

3. How small is the smallest seahorse?

The Pigmy Seahorse averages ½ inch in length, which is smaller than an average human fingernail.

4. Would you rather be an Indo-Pacific Sailfish or a Pigmy Seahorse? Why?

## QUESTIONS

1. HOW MANY GALLONS OF GASOLINE DOES THE AVERAGE CAR HOLD?

NAME THE MINIMUM AND THE MAXIMUM NUMBER OF GALLONS.

THE AVERAGE FUEL TANK CAPACITY OF A CAR IS BETWEEN 10.5 AND 18.5 GALLONS OF GASOLINE. (+10.5 TO +18.5 GALLONS OF GASOLINE)

2.IN THE FUTURE WHAT TYPE OF VEHICLE WOULD YOU LIKE TO OWN AND DRIVE?

# POSITIVE NUMBERS

WHEN YOU NEED GASOLINE IN YOUR VEHICLE, YOU DRIVE TO A GASOLINE STATION TO FILL UP WITH ONE OF THE BRANDS OF GASOLINE. THAT NUMBER REPRESENTS A POSITIVE NUMBER OF GALLONS PUMPED INTO YOUR VEHICLE.

# ZERO

ZERO WOULD REPRESENT NO GALLONS OF GASOLINE IN THE VEHICLE.

# NEGATIVE NUMBERS

AS YOU DRIVE YOUR VEHICLE, THE NUMBER OF GALLONS OF GASOLINE IN YOUR CAR DECREASES. THIS DECREASING NUMBER REPRESENTS A NEGATIVE NUMBER OF GALLONS USED BY YOUR VEHICLE.

3. HOW MANY MILES PER GALLON DO THE 2024 CARS AVERAGE?

ACCORDING TO THE EPA (ENVIRONMENTAL PROTECTION AGENCY), THE AVERAGE 2024 VEHICLE GETS 28 MILES PER GALLON.

THE NUMBER IS BEING PUSHED HIGHER AS FUEL ECONOMY IMPROVES.

# POSITIVE NUMBER

YOUR FAMILY GOES TO THE BEACH DURING THE SUMMER. WHAT SHAPE WOULD YOU LIKE TO BUILD ON THE BEACH WITH SAND AND YOUR UTENSILS?

LET'S BUILD A SANDCASTLE 3 FEET TALL.

# ZERO

ZERO WOULD REPRESENT THE FLAT LEVEL OF THE SAND WHERE YOU ARE WALKING AT THE BEACH.

# NEGATIVE NUMBER

TO STAY PUT IN THE SAND, A SAND CRAB BURROWS DOWN QUICKLY AND OFTEN. A SAND CRAB MOVES ONLY BACKWARD, WHICH REPRESENTS A NEGATIVE NUMBER.

# QUESTIONS

1. WHAT OTHER SHAPE WOULD YOU LIKE TO BUILD WITH THE SAND AND UTENSILS ON THE BEACH?

2. WHY DOES THE SAND CRAB MOVE BACKWARD? A SAND CRAB MOVES ONLY BACKWARD BECAUSE IT HAS NO CLAWS ON ITS FIRST PAIR OF LEGS.

THEY SWIM AND BURROW, MOVING BACKWARDS, AND CONSTANTLY REBURY THEMSELVES AS THEY FOLLOW THE WAVES.

THE CRABS FACE SEAWARD WITH ONLY THEIR EYES AND THE FIRST ANTENNA SHOWING. THEN THE SAND CRABS UNCOIL A SECOND PAIR OF FEATHERLIKE ANTENNAE AND SWEEP THEM THROUGH THE WATER TO FILTER OUT TINY PLANKTON, WHICH IS THEIR FOOD.

THREE FEET TALL IS A POSITIVE NUMBER.
POSITIVE 3 = +3 FEET TALL.

3. WHAT OTHER ANIMALS ARE KNOWN TO WALK OR MOVE BACKWARD?

OCTOPUSES MOVE BACKWARD BY USING THEIR TENTACLES AND JET PROPULSION.

BIRDS LIKE THE NORTHERN FLICKER CLIMB DOWN TREES IN A BACKWARD MOTION.

WALKING BACKWARD SERVES SPECIFIC PURPOSES RELATED TO FEEDING, ESCAPING PREDATORS OR GENERAL NAVIGATION.

## QUESTIONS

1. WHAT IS YOUR FAVORITE CLASS IN SCHOOL? WHAT IS YOUR LEAST FAVORITE CLASS IN SCHOOL?

2. WHAT TESTS AND QUIZZES DO YOU LIKE TO TAKE? WHAT TESTS AND QUIZZES DO YOU NOT LIKE TO TAKE?

3. WOULD YOU LIKE TO WORK AS A TEACHER, COUNSELOR, VICE-PRINCIPAL, PRINCIPAL, CAFETERIA WORKER OR CUSTODIAN IN THE FUTURE? WHY?

95

# POSITIVE NUMBERS

AT SCHOOL STUDENTS TAKE TESTS AND QUIZZES. SUPPOSE A STUDENT RECEIVES POINTS FOR CORRECT ANSWERS ON A TEST WHICH ARE ADDED TOGETHER FOR A GRADE OF 95.

THESE 95 POINTS ARE A POSITIVE NUMBER. 95 POINTS REPRESENTS +95 POINTS.

## ZERO

ZERO IS THE NUMBER OF POINTS YOU HAVE ON THE QUIZ OR TEST BEFORE THE TEACHER ADDS UP THE NUMBER OF POSITIVE POINTS TO YOUR QUIZ OR TEST, WHICH IS 95 POINTS.

## NEGATIVE NUMBERS

ON THE SAME TEST, THE STUDENT HAD 5 POINTS SUBTRACTED FROM THE NUMBER OF POINTS AVAILABLE FOR A 100 GRADE.

FIVE POINTS SUBTRACTED FROM THE TOTAL NUMBER OF POINTS AVAILABLE IS
100 POINTS - 5 POINTS = 95 POINTS.

FIVE POINTS SUBTRACTED IS A NEGATIVE NUMBER WHICH IS -5 POINTS.

THE FINAL GRADE IS A 95 FOR THE STUDENT.

# POSITIVE NUMBER

A POSITIVE NUMBER WOULD REPRESENT THE NUMBER OF FEET THE PLANE ASCENDS INTO THE SKY AFTER IT TAKES OFF FROM THE AIRPORT HEADING TO ITS NEXT DESTINATION.

COMMERCIAL AIRLINES CRUISE BETWEEN POSITIVE 30,000 FEET AND POSITIVE 42,000 FEET. (+30,000 AND +42,000 FEET)

## ZERO

AN AIRPLANE IS SITTING ON THE GROUND AT THE AIRPORT GATE WHILE WORKERS ARE LOADING LUGGAGE, FUEL, FOOD, DRINKS, AND PASSENGERS INSIDE THE AIRPLANE.

THE PLANE IS ON THE GROUND WHILE LOADING AT THE GATE AND CRUISING ON THE RUNWAY BEFORE IT TAKES OFF FROM THE AIRPORT. ZERO FEET INDICATES THE PLANE IS SITTING ON THE GROUND.

# +30,000 AND +42,000 FT

## NEGATIVE NUMBERS

A NEGATIVE NUMBER WOULD REPRESENT THE NUMBER OF FEET THE PLANE DESCENDS TO LAND AT THE NEXT AIRPORT FOR THIS TRIP. THE PLANE WOULD DESCEND BETWEEN NEGATIVE 30,000 AND NEGATIVE 42,000 FEET. (-30,000 AND -42,000 FEET)

## QUESTIONS

1. WHAT DESTINATION WOULD YOU LIKE TO FLY TO FOR A VACATION?

2. WHO WOULD GO WITH YOU ON YOUR VACATION?

3. WHEN YOU GROW UP WOULD YOU BE A FLIGHT ATTENDANT OR BE A PILOT?

## QUESTIONS

1. WHAT COUNTRY GAVE THE STATUE OF LIBERTY TO THE UNITED STATES OF AMERICA? WHY?

ON JULY 4, 1884, FRANCE GIFTED THE STATUE OF LIBERTY TO THE UNITED STATES AS A SYMBOL OF FRIENDSHIP AND INTERNATIONAL COOPERATION DURING THE AMERICAN REVOLUTION.

THE STATUE WAS A COMMEMORATION OF THE 100TH ANNIVERSARY OF THE U.S. DECLARATION OF INDEPENDENCE AND THE SUCCESS OF AMERICAN DEMOCRACY.

# POSITIVE NUMBERS

THE STATUE OF LIBERTY IS LOCATED NEAR NEW YORK CITY HARBOR. VISITORS CLIMB 354 STEPS (22 STORIES) TO LOOK OUT FROM 25 WINDOWS IN THE CROWN OF THE STATUE. CLIMBING UP 354 STEPS IS A POSITIVE NUMBER.

## ZERO

ZERO IS GROUND LEVEL OUTSIDE OF THE STATUE OF LIBERTY BEFORE YOU CLIMB THE STAIRS TO THE TOP.

## NEGATIVE NUMBERS

DESCENDING 354 STEPS IS A NEGATIVE NUMBER WHICH IS -354 STEPS.

2.HOW MANY REPLICAS OF THE STATUE OF LIBERTY ARE LOCATED AROUND PARIS?

AT LEAST 5 REPLICAS OF THE STATUE OF LIBERTY ARE LOCATED AROUND PARIS.

3. WHAT IS IN THE STATUE OF LIBERTY'S LEFT HAND?

THE STATUE OF LIBERTY IS HOLDING A TABLET WHERE THE DATE OF THE DECLARATION OF INDEPENDENCE JULY IV MDCCLXXVI (1776) IS WRITTEN.

# POSITIVE NUMBERS

YOUR SISTER IS 2 YEARS OLDER THAN YOU. TWO YEARS OLDER REPRESENTS A POSITIVE NUMBER 2 OR +2.

## ZERO

IF YOU HAVE ZERO YEARS BETWEEN YOU AND ONE OR TWO OF YOUR SIBLINGS, YOU COULD HAVE A TWIN OR YOU COULD BE ONE OUT OF THREE BABIES WHO ARE TRIPLETS.

## NEGATIVE NUMBERS

YOUR BROTHER IS 3 YEARS YOUNGER THAN YOU.
THREE YEARS YOUNGER REPRESENTS A NEGATIVE NUMBER 3 OR -3.

## QUESTIONS

1. HOW MANY SIBLINGS DO YOU HAVE IN YOUR FAMILY? ARE THEY OLDER OR YOUNGER?

2. IF YOU HAD A CHOICE, WOULD YOU RATHER BE THE OLDEST, MIDDLE OR YOUNGEST CHILD IN A FAMILY? WHY?

# POSITIVE NUMBERS

THE LINCOLN MEMORIAL, HONORING THE 16TH PRESIDENT IN WASHINGTON, D. C. HAS 58 STEPS GOING INTO THE MEMORIAL.

FIFTY-EIGHT STEPS REPRESENT A POSITIVE 58 GOING UP INTO THE MEMORIAL WHICH IS +58 STEPS.

THERE ARE 87 STEPS FROM THE CHAMBER TO THE REFLECTING POOL WHICH IS +87 STEPS.

## ZERO

BEFORE YOU ENTER THE LINCOLN MEMORIAL AND GO UP THE STEPS, YOU WOULD BE AT GROUND LEVEL WHICH IS AT ZERO STEPS.

# NEGATIVE NUMBER

## EXITING 58 STEPS FROM THE LINCOLN MEMORIAL IS A NEGATIVE 58 = -58 STEPS

# QUESTIONS

1. WHAT WAS PRESIDENT ABRAHAM LINCOLN FAMOUS FOR ISSUING IN 1863? PRESIDENT LINCOLN ISSUED THE EMANCIPATION PROCLAMATION THAT DECLARED FOREVER FREE THOSE SLAVES WITHIN THE CONFEDERACY.

2. THE ARCHITECT, HENRY BACON, MODELED THE LINCOLN MEMORIAL AFTER WHAT BUILDING IN ATHENS, GREECE? WHY? HENRY BACON MODELED THE LINCOLN MEMORIAL AFTER THE PARTHENON.

HE FELT A MEMORIAL DEDICATED TO A MAN WHO DEFENDED DEMOCRACY SHOULD ECHO THE BIRTHPLACE OF DEMOCRACY.

3. WHAT DO THE 36 COLUMNS SIGNIFY IN THE LINCOLN MEMORIAL? THE 36 COLUMNS CORRESPOND TO THE NUMBER OF STATES IN THE UNION DURING LINCOLN'S LIFETIME.

4. WHAT RELATIVE OF LINCOLN'S WAS PRESENT TO OBSERVE THE LINCOLN MEMORIAL OFFICIAL DEDICATION IN MAY 1922?

ROBERT TODD LINCOLN, THE ONLY SURVIVING SON OF THE FORMER PRESIDENT OBSERVED THE OFFICIAL DEDICATION IN MAY 1922.

## QUESTIONS

1. HOW LONG DO YOU BOIL WATER TO COOK EGGS IN THE SHELL?

FIFTEEN MINUTES FROM START TO FINISH. LET THE EGGS COOL BEFORE REMOVING THE SHELL.

2. WHAT FOOD DO YOU LIKE TO EAT WITH EGGS IN IT?

SCRAMBLED EGGS, QUICHE, CAKES, COOKIES AND BREADS.

# POSITIVE NUMBERS

THE BOILING POINT OF WATER IS 212 DEGREES FAHRENHEIT WHICH IS A POSITIVE 212 DEGREES, +212 DEGREES. WHEN A LIQUID REACHES ITS BOILING POINT, IT RAPIDLY TRANSITIONS FROM A LIQUID TO A GAS OR VAPOR.

## ZERO

ZERO MINUTES WOULD BE THE TIME BEFORE YOU START TO BOIL THE EGG AND COOK IT.

## NEGATIVE NUMBERS

WATER FREEZES AT 32 DEGREES FAHRENHEIT. THE FREEZING POINT IS THE TEMPERATURE BY WHICH A LIQUID TURNS INTO A SOLID.

3. WHAT IS THE DIFFERENCE IN TEMPERATURE BETWEEN THE BOILING POINT AND THE FREEZING POINT OF WATER?

180 DEGREES IS THE DIFFERENCE BETWEEN 212 DEGREES AND 32 DEGREES FAHRENHEIT, 212 DEGREES - 180 DEGREES = 32 DEGREES FAHRENHEIT.

SUBTRACTING 180 DEGREES IS THE SAME AS NEGATIVE 180 DEGREES.

# POSITIVE NUMBER

IN 1931 PRESIDENT HOOVER PRESSED A BUTTON OPENING THE EMPIRE STATE BUILDING IN NEW YORK CITYAND TURNING ON THE LIGHTS FOR THE VERY FIRST TIME.

IT WAS THE WORLD'S TALLEST BUILDING IN THE WORLD WITH A 102 FLOOR OBSERVATION DECK. PEOPLE FROM ALL OVER THE WORLD PAID TEN CENTS TO PEER THROUGH A TELESCOPE. TODAY, IT COST $50.00 TO RIDE TO THE OBSERVATION DECK.

102 STORIES ABOVE GROUND REPRESENTS POSITIVE 102 OR +102 STORIES.

## ZERO

GROUND LEVEL ON THE SIDEWALK WOULD REPRESENT ZERO FEET.

## NEGATIVE NUMBER

RIDING DOWN THE ELEVATOR FROM THE 102ND FLOOR WILL BE -102 FLOORS WHICH TAKES ABOUT A MINUTE.

# QUESTIONS

1. WOULD YOU CHOOSE TO RIDE THE ELEVATOR OR CLIMB THE STEPS TO THE OBSERVATION TOWER? WHY?

CLIMBING THE STEPS IS FREE BUT THERE IS A CHARGE FOR USING THE ELEVATOR TO REACH THE 86TH AND 102ND FLOOR. TODAY THE PRICE FOR THE EXPRESS PASS FOR ALL VISITORS TO THE 86TH AND 102ND FLOORS IS $120 PER PERSON.

2. WHAT FAMOUS MOVIE WAS MADE IN 1933 THAT FEATURES THE EMPIRE STATE BUILDING THAT'S CONSIDERED ONE OF THE MOST ICONIC MOMENTS IN MOVIE HISTORY? "KING KONG"

3. HOW LONG DID IT TAKE TO CONSTRUCT THE EMPIRE STATE BUILDING?

IT WAS COMPLETED IN RECORD TIME, TAKING ONLY 20 MONTHS FROM START TO FINISH. THE BUILDING WAS COMPLETED AHEAD OF SCHEDULE AND UNDER BUDGET.

4. THE SHAPE OF THE EMPIRE STATE BUILDING WAS THOUGHT TO BE INSPIRED BY WHAT OBJECT IN YOUR CLASSROOM?

THE DISTINCTIVE SHAPE OF THE ESB WITH ITS LONG STRAIGHT TOWER AND SHARP POINTED SPIRE AT THE TIP IS THOUGHT TO BE MODELLED ON THE SHAPE OF A PENCIL.

## QUESTIONS

1, WHAT ARE 2 OTHER RULES FOR DRIVERS IN A SCHOOL ZONE UNLESS THE CELL PHONE IS SET UP AS A HANDS-FREE DEVICE?

THE USE OF CELL PHONES IN SCHOOL ZONES IS PROHIBITED. STOP BOTH WAYS FOR SCHOOL BUSES THAT ARE LOADING AND UNLOADING CHILDREN.

2. DO YOU THINK SEAT BELTS SHOULD BE INSTALLED IN SCHOOL BUSES?

3. WHAT IS THE FINE FOR SPEEDING IN A SCHOOL ZONE IN TEXAS?

YOU PAY FROM $184 TO $334 OR MORE DEPENDING ON HOW FAST YOU WERE DRIVING.

# POSITIVE NUMBERS

THE STREET IN FRONT OF THE ELEMENTARY SCHOOL HAS A SPEED LIMIT FOR VEHICLES OF 35 MILES PER HOUR WHEN THE SCHOOL IS NOT IN SESSION FOR STUDENTS.
POSITIVE 35 MPH = +35 MPH

35

## ZERO

WHEN VEHICLES STOP AT A STOP SIGN OR A RED LIGHT, THEY ARE GOING ZERO MILES PER HOUR.

## NEGATIVE NUMBERS

WHEN THE STUDENTS ARE COMING TO SCHOOL AT THE BEGINNING OF THE DAY OR LEAVING SCHOOL AT THE END OF THE DAY, THE SPEED LIMIT CHANGES TO 20 MILES PER HOUR FOR ALL VEHICLES.

THE DRIVERS NEED TO SLOW DOWN AND DECREASE THE SPEED OF THEIR VEHICLE BY 15 MILES PER HOUR OR POSSIBLY GET A TRAFFIC TICKET FROM THE POLICE.

DECREASING THE SPEED OF THE VEHICLE BY 15 MPH IS A NEGATIVE NUMBER, WHICH IS -15 MPH.

# POSITIVE NUMBER

A PERSON HAS A NORMAL BODY TEMPERATURE OF 98.7 DEGREES FAHRENHEIT WHICH IS A POSITIVE 98.7DEGREES, +98.7 DEGREES.

## NEGATIVE NUMBER

A BODY TEMPERATURE USUALLY BETWEEN 100.4 AND 102.2 DEGREES IS CONSIDERED A LOW-GRADE TEMPERATURE.

102.2 DEGREES - 98.7 DEGREES = 3.5 DEGREES.

TO BRING DOWN A LOW-GRADE FEVER TO A NORMAL TEMPERATURE, ONE WOULD NEED TO DECREASE THE TEMPERATURE BY 3.5 DEGREES, WHICH IS A NEGATIVE NUMBER 3.5 DEGREES = -3.5 DEGREES.

# QUESTIONS

1. A LOW-GRADE FEVER USUALLY DOES NOT REQUIRE TREATMENT UNLESS THE FEVER OCCURS IN WHAT TYPE OF HUMAN? WHY?

A VERY YOUNG INFANT. FEVERS IN INFANTS CAN INDICATE A SERIOUS ILLNESS.

2. HOW DO YOU BRING DOWN A FEVER WITHOUT MEDICATION?

YOU MAY WANT TO BATHE IN LUKEWARM WATER. USE A FAN TO COOL OFF. PLACE A COOL COMPRESS ON FOREHEAD, NECK, FEET AND HANDS. THIS IS WHERE A LOT OF HEAT IS RELEASED. STAY HYDRATED BY DRINKING PLENTY OF FLUIDS.

3. WHEN DO YOU GO TO THE EMERGENCY ROOM WITH A FEVER?

IF THE TEMPERATURE IS ABOVE 102.2 DEGREES FAHRENHEIT AND ESPECIALLY IF YOU HAVE OTHER SYMPTOMS SUCH AS DIFFICULTY BREATHING, WAKING UP OR YOU ARE UNABLE TO KEEP FLUIDS DOWN.

102.2 DEGREES IS A POSITIVE NUMBER = + 102.2 DEGREES.

## ZERO

ZERO WOULD MEAN THE ANIMAL HAS NO ARMS.

## NEGATIVE NUMBER

IF A PREDATOR, SUCH AS A SHARK, GRABS AN OCTOPUS'S ARM, THE OCTOPUS CAN SHED ITS LIMB AND SWIM AWAY.

NEGATIVE NUMBER 1 = -1 LEG.

## QUESTIONS

1. HOW LONG DOES IT TAKE THE OCTOPUS TO REBUILD ITS ARM?

THE ANSWER IS 130 DAYS OR MORE THAN 4 MONTHS TO REBUILD ITS ARM.

2. WHAT IS THE RECORD FOR THE GREATEST NUMBER OF LEGS ON AN OCTOPUS?

THE OCTOPUS WAS CAUGHT IN 1998 IN JAPAN WITH 96 LEGS.

# POSITIVE NUMBERS

AN OCTOPUS IS A SEA ANIMAL WITH SOFT, ROUNDED BODIES, LARGE EYES, AND EIGHT LONGARMS OR TENTACLES. OCTOPUSES ARE CARNIVOROUS PREDATORS THAT EAT CRABS,
LOBSTERS, WORMS, MOLLUSKS AND SMALL FISH.

OCTOPUSES HAVE SEVERAL DEFENSE MECHANISMS, INCLUDING SHOOTING A JET OF WATER BACKWARD TO MOVE QUICKLY AWAY FROM DANGER.

RELEASING AN INKY FLUID TO DARKEN THE WATER AND CONFUSE ENEMIES, AND BEING BONELESS SO THEY CAN SQUEEZE INTO TIGHT SPACES.

POSITIVE NUMBER 8 = +8 ARMS.

3. WHY DID THE OCTOPUS HAVE 96 LEGS?

EACH OF ITS EIGHT NORMAL TENTACLES HAD BRANCHED TO PRODUCE ADDITIONAL ONES, YIELDING AN AMAZING 96 TENTACLES IN TOTAL.

4. WHAT ANIMALS HAVE NO ARMS OR LEGS?

SALAMANDERS, SNAKES, LEGLESS LIZARDS, EARTHWORMS, SNAILS AND FISH.

## ZERO

ZERO WOULD REPRESENT THE AMOUNT OF MONEY YOU RECEIVED FROM YOUR GUESTS BEFORE YOUR BIRTHDAY WHICH IS $0.00.

## QUESTIONS

1. WOULD YOU PREFER TO SPEND YOUR MONEY NOW TO BUY SOMETHING FOR YOUR BIRTHDAY OR SAVE YOUR BIRTHDAY MONEY AND SPEND IT LATER?

2. WHAT ITEM(S) WOULD YOU LIKE TO BUY WITH YOUR BIRTHDAY MONEY?

3. WHAT THEME WOULD YOU CHOOSE FOR YOUR NEXT BIRTHDAY CELEBRATION?

# POSITIVE NUMBERS

SUPPOSE YOU TURNED 6 YEARS OLD ON YOUR BIRTHDAY AND YOU RECEIVED \$6 AND \$7 FROM TWO GUESTS AT YOUR BIRTHDAY PARTY.

HOW MUCH MONEY DID YOU RECEIVE AT YOUR BIRTHDAY PARTY?

\$6 + \$7 = \$13. DOES \$13 REPRESENT A POSITIVE NUMBER OR A NEGATIVE NUMBER?

RECEIVING \$13 MEANS THAT 13 IS A POSITIVE NUMBER.

YOU HAVE \$13 THAT YOU CAN SPEND ON YOURSELF OR SAVE FOR SOMETHING YOU WANT IN THE FUTURE. BOTH \$6 RECEIVED AND \$7 RECEIVED ARE POSITIVE NUMBERS.

## NEGATIVE NUMBERS

YOUR MOTHER AND/OR FATHER ACCOMPANIES YOU TO THE STORE TO PURCHASE 1 OR MORE BIRTHDAY PRESENTS FOR \$13 OR LESS.

SUPPOSE YOU BUY 2 ITEMS AT THE STORE THAT COST A TOTAL OF \$10. WOULD \$10 BE A POSITIVE OR NEGATIVE NUMBER?

COSTING \$10 WOULD BE A NEGATIVE NUMBER, WHICH IS - \$10. THE ITEMS COST \$10, SO YOU WOULD GIVE THE CASHIER AT THE STORE \$10 PLUS SALES TAX TO BUY THE TWO ITEMS AND TAKE THEM HOME. SPENDING \$10 IS A NEGATIVE NUMBER.

\$13 - \$10 = \$3 WHICH IS A POSITIVE NUMBER.

# POSITIVE NUMBERS

THE CHEETAH IS THE FASTEST LAND ANIMAL IN THE WORLD WITH A TOP SPEED OF 75 MILES PER HOUR.

IN 3 SECONDS, A CHEETAH CAN ACCELERATE FROM 0 TO 60 MILES PER HOUR.

## ZERO

ZERO SPEED WOULD REPRESENT ZERO MILES PER HOUR. THE ANIMALS WOULD BE STANDING STILL AND NOT MOVING.

## NEGATIVE NUMBERS

THE THREE-TOED SLOTH IS THE SLOWEST MOVING MAMMAL ON EARTH. IT CLIMBS AT A SPEED OF 6-8 FEET PER MINUTE. WHEN IS DESCENDS FROM THE TREE, THE SPEED WOULD BE A NEGATIVE 6-8 FEET PER MINUTE.

# QUESTIONS

1. WHAT ARE THE FOUR REASONS THE CHEETAH IS THE FASTEST LAND ANIMAL IN THE WORLD?

THE CHEETAHS HAVE A LIGHT BODY WEIGHT. THEY ARE ALWAYS READY TO PROVIDE POWERFUL TRACTION TO THE GROUND. TO ACCELERATE, ONLY ONE FOOT AT A TIME IS IN CONTACT WITH THE GROUND. CHEETAHS HAVE SEMI-RETRACTABLE CLAWS WHICH NEVER FULLY RETRACT.

2. WHY ARE THE SLOTHS THE SLOWEST MOVING ANIMAL?

THE SLOWNESS OF SLOTHS IS ATTRIBUTABLE TO THEIR DIET. THEY DEPEND ON A DIET OF LEAVES, WHICH ARE POOR IN NUTRIENTS AND LOW IN CALORIES. THEIR SLOW-MOVING LIFESTYLE, WHICH FAVORS A SLOW METABOLISM, IS DESIGNED TO CONSERVE ENERGY.

SLOTHS HAVE THE SLOWEST DIGESTION RATE OF ANY MAMMAL, WITH A SINGLE LEAF TAKING UP TO 30 DAYS TO PASS THROUGH THEIR DIGESTIVE TRACT.

3. HOW MANY CALORIES A DAY DO THE SLOTHS EAT?

THEY ONLY EAT 160 CALORIES A DAY SO THEY CONSERVE THEIR ENERGY.

4. WOULD YOU RATHER BE A CHEETAH OR A SLOTH? WHY?

## QUESTIONS

1. WHAT BOARD GAMES DO YOU LIKE TO PLAY?

2. WHICH DIRECTION DO YOU MOVE TO WIN THE GAME? YOU MOVE IN A POSITIVE DIRECTION TO WIN THE GAME.

3. DO YOU WANT TO MAKE UP YOUR OWN BOARD GAME?

# POSITIVE NUMBERS

DO YOU LIKE TO PLAY BOARD GAMES SUCH AS MONOPOLY, CHECKERS OR CHESS?

WHEN YOU ROLL THE DICE AND YOU MAY MOVE FORWARD 6 SPACES, DOES THAT NUMBER SIGNIFY A POSITIVE OR NEGATIVE NUMBER?

MOVING FORWARD SIGNIFIES A POSITIVE NUMBER WHICH IS POSITIVE 6 OR +6.

## ZERO

ZERO MOVES ON A BOARD GAME MEANS YOU ARE NOT MOVING FORWARD OR BACKWARD.

## NEGATIVE NUMBERS

IF YOU SLIDE DOWN A LADDER SO YOUR TOKEN IS NOW CLOSER TO WHERE YOUR TOKEN STARTED THE GAME, YOUR TOKEN IS RETURNED TO THE HOME SPACE, OR YOU PICK A CARD AND YOUR TOKEN GOES BACKWARD TOWARD THE START POSITION, THEN YOU ARE MOVING IN A POSITIVE OR NEGATIVE DIRECTIONS?
ANSWER: A NEGATIVE DIRECTION.

# POSITIVE NUMBER

A PENTAGON IS A POLYGON WHICH HAS 5 STRAIGHT SIDES AND FIVE ANGLES.

NAME A POLYGON WHICH HAS 6 STRAIGHT SIDES AND 6 ANGLES.

ANSWER: HEXAGON
THE HEXAGON HAS ONE MORE SIDE THAN A PENTAGON.

ONE MORE SIDE INDICATES A POSITIVE ONE. + ONE = ONE MORE SIDE.

# ZERO

IF A POLYGON HAS ZERO SIDES, THEN THE POLYGON DOES NOT EXIST.

# NEGATIVE NUMBER

PLEASE DRAW A 4-SIDED FIGURE: ONE DIMENSIONAL SQUARE.

PLEASE DRAW A 3-SIDED FIGURE: ONE DIMENSIONAL TRIANGLE.

THE TRIANGLE HAS ONE LESS SIDE THAN A SQUARE. ONE LESS SIDE INDICATES A NEGATIVE ONE. (-1)

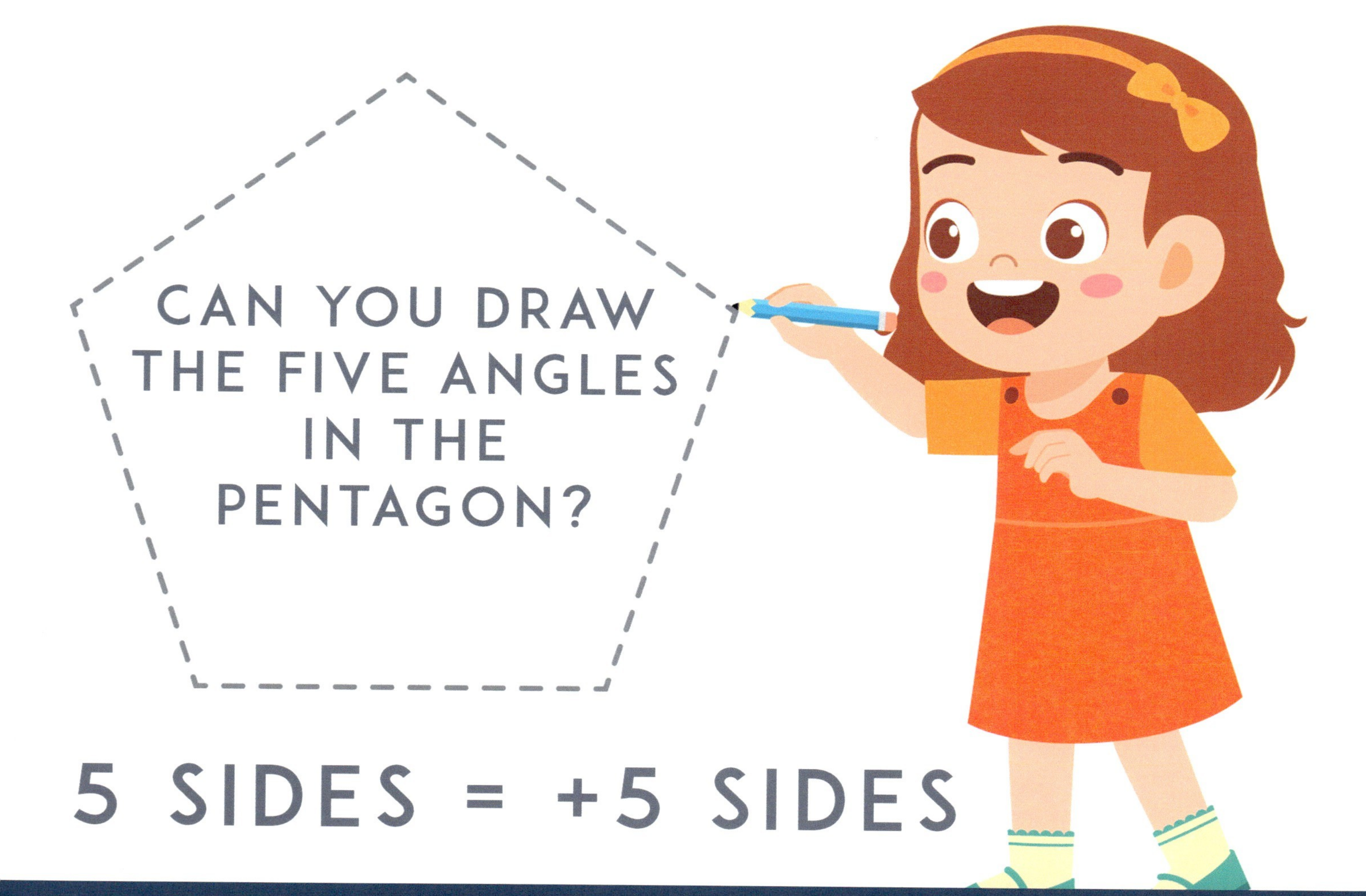


# 5 SIDES = +5 SIDES

## QUESTIONS

1. A SQUARE IS ONE TYPE OF POLYGON WITH 4 STRAIGHT SIDES AND 4 ANGLES.

NAME 3 OTHER TYPES OF POLYGONS WITH 4 STRAIGHT SIDES AND 4 ANGLES.
RECTANGLES, TRAPEZOIDS, AND KITES HAVE 4 STRAIGHT SIDES AND 4 ANGLES.

2. THERE ARE 3 TYPES OF TRIANGLES WITH 3 STRAIGHT LINES AND 3 ANGLES.

NAME THE 3 TYPES OF TRIANGLES.
EQUILATERAL TRIANGLES,(3 SIDES EQUAL), ISOSCELES TRIANGLES, (2 SIDES EQUAL) AND SCALENE TRIANGLES, (NO SIDES EQUAL) HAVE 3 STRAIGHT LINES AND 3 ANGLES.

3. DO YOU KNOW THE NAMES OF OTHER POLYGONS?
OTHER NAMES FOR POLYGONS ARE THE HEPTAGON, (7 SIDES), OCTAGON, (8 SIDES), NONAGON, (9 SIDES), AND DECAGON, (10 SIDES).

## QUESTIONS

1. WHAT FISH DO YOU LIKE TO CATCH?

2. WHAT FISH DO YOU LIKE TO EAT?

3. WHERE DO YOU PREFER TO FISH?

4. WHERE IS THE ONLY LIVING CORAL BARRIER REEF IN NORTH AMERICA LOCATED?

THE FLORIDA KEYS ARE HOME TO NORTH AMERICA'S ONLY CORAL BARRIER REEF AND MORE THAN 40 SPECIES OF REEF-BUILDING CORAL. IT EXTENDS FOR 350 MILES. THE TOTAL AREA IS NEARLY THREE TIMES THE SIZE OF YELLOWSTONE NATIONAL PARK.

# POSITIVE NUMBERS

YOU LIKE TO GO FISHING AND YOU ASK YOUR FRIEND TO GO FISHING WITH YOU.

WHEN YOU AND YOUR FRIEND ARE IN THE BOAT FISHING, YOU SEE A LIGHTHOUSE ON THE SHORE.

THE LIGHTHOUSE IS 150 FEET TALL WHICH REPRESENTS A POSITIVE 150 FEET ABOVE SEA LEVEL.
POSITIVE 150 FEET = + 150 FEET

## ZERO

WE USE THE TERM 'SEA LEVEL' TO HELP DESCRIBE LANDFORMS THAT ARE ABOVE THE WATER, LIKE LIGHTHOUSES.

WE MAY ALSO DESCRIBE FORMS BELOW SEA LEVEL LIKE THE ONLY LIVING CORAL BARRIER REEF IN NORTH AMERICA LOCATED IN THE FLORIDA KEYS
10-15 FEET DEEP.

(-10 TO -15 FEET BELOW SEA LEVEL)

## NEGATIVE NUMBERS

YOU AND YOUR FRIEND ARE CATCHING A LOT OF FISH, SO YOU DROP YOUR ANCHOR AND STAY IN THIS LOCATION TO FISH. THE BOYS ARE DROPPING AN ANCHOR 100 FEET.

THE ANCHOR LINE IS 100 FEET LONG WHICH REPRESENTS NEGATIVE 100 FEET BELOW SEA LEVEL.

{-100 FEET)= 100 FEET BELOW SEA LEVEL.

# POSITIVE NUMBER

SUNFLOWERS GROW QUICKLY AND THEIR HEIGHT RANGES FROM 1 FOOT TO 17 FEET TALL.

(+1 TO +17 FEET TALL)

## ZERO

GROUND LEVEL MEANS THERE IS GROUND BELOW TO PLANT FLOWERS AND VEGETABLE SEEDS AND THERE IS SPACE ABOVE NEEDED FOR AIR AND RAIN FOR THE PLANTS TO GROW AND PRODUCE FLOWERS AND VEGETABLES.

## NEGATIVE NUMBER

WHEN PLANTING A FLOWER BED OR A VEGETABLE GARDEN, PLACE THE SEED 1-2 INCHES BELOW GROUND LEVEL AND WATER DURING THE CORRECT SEASON FOR PLANTING.

(-1 TO - 2 INCHES BELOW GROUND LEVEL)

## QUESTIONS

1. WHAT IS THE QUICKEST VEGETABLE TO GROW FROM SEED?

RADISHES ARE READY TO EAT AS SOON AS THREE WEEKS AFTER PLANTING THE SEEDS. WHAT ARE YOUR FAVORITE VEGETABLES, FRUITS, OR BERRIES TO GROW IN YOUR GARDEN?

2. HOW MANY SPECIES OF SUNFLOWERS ARE THERE IN THE WORLD?

THERE ARE 70 SPECIES OF SUNFLOWERS IN THE WORLD AND MOST ARE NATIVE TO NORTH AMERICA.

3. WHAT IS THE LARGEST SPECIES OF SUNFLOWER?

SUNFLOWER 'GIRAFFE" REACHES A HEIGHT OF 17 FEET.

4. WHAT STATE HAS THE MOST SUNFLOWERS?

KANSAS IS THE SUNFLOWER STATE BECAUSE THEIR WEATHER IS PERFECTLY SUITED FOR THEM.

5. WHAT DO SUNFLOWERS REPRESENT?

A LONG LIFE AND LASTING HAPPINESS.

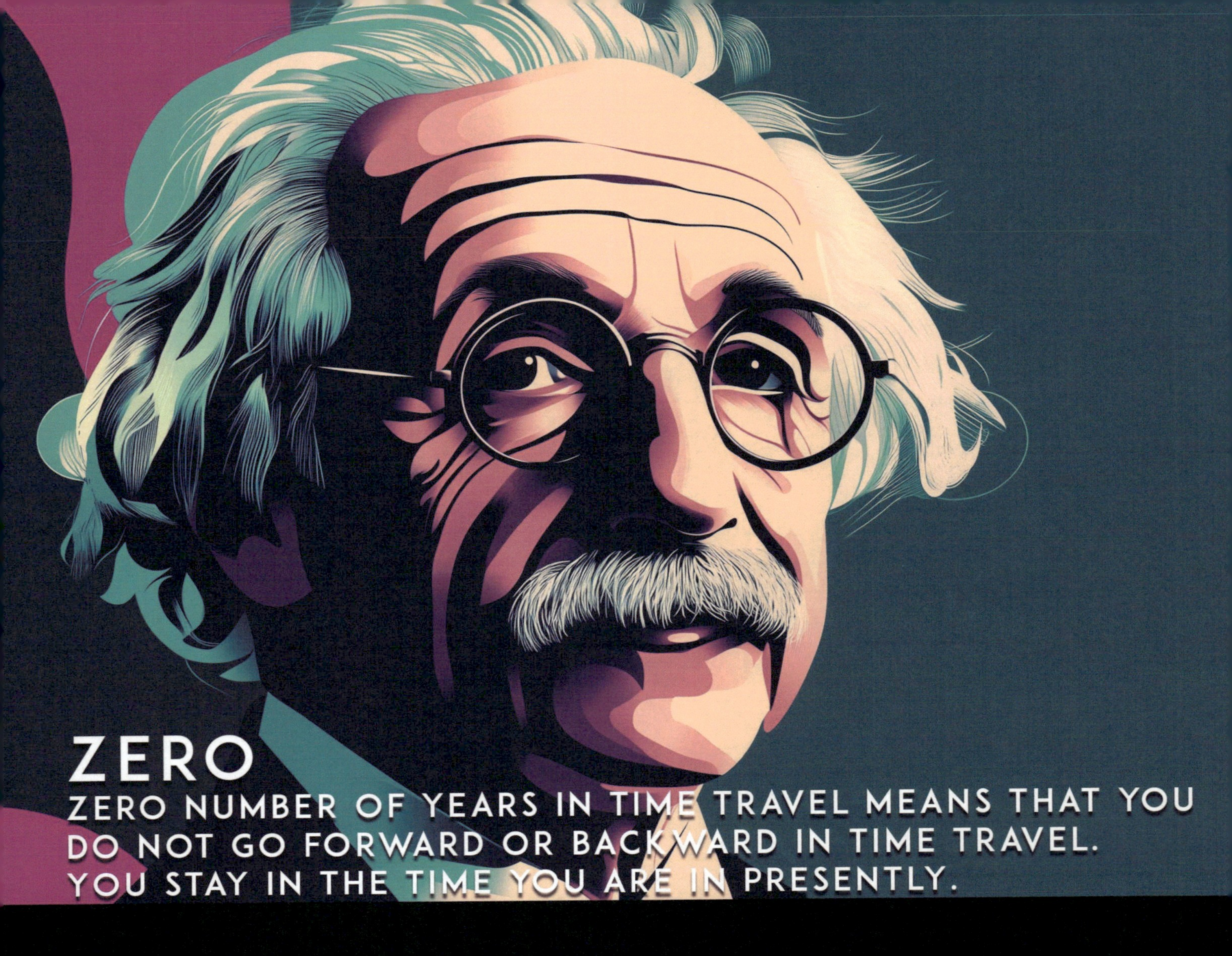

## ZERO

ZERO NUMBER OF YEARS IN TIME TRAVEL MEANS THAT YOU DO NOT GO FORWARD OR BACKWARD IN TIME TRAVEL. YOU STAY IN THE TIME YOU ARE IN PRESENTLY.

## NEGATIVE NUMBER

TRAVELING INTO THE PAST IS A NEGATIVE NUMBER.

IF YOU WANTED TO SEE THE DINOSAURS, YOU WOULD TRAVEL BACK IN TIME 66 MILLION YEARS TO 252 MILLION YEARS AGO.

BOTH 66 MILLION AND 252 MILLION YEARS AGO WOULD BE NEGATIVE NUMBERS.

(-66 MILLION TO -252 MILLION YEARS AGO)

# POSITIVE NUMBERS

TIME TRAVEL IS THE IMAGINED ABILITY TO TRAVEL TO THE PAST OR FUTURE. IT IS A COMMON CONCEPT IN FICTION, ESPECIALLY SCIENCE FICTION, WHERE TIME TRAVEL IS OFTEN ACHIEVED USING A TIME MACHINE.

TRAVELING INTO THE FUTURE IS A POSITIVE NUMBER.

H. G. WELLS WROTE "THE TIME MACHINE" IN WHICH OFFERS A VISION OF HUMANITY'S FUTURE. A SCIENTIST BUILDS A TIME MACHINE AND TRAVELS TO THE FUTURE. HE TRAVELS 800,806 YEARS INTO THE FUTURE, WHICH IS A POSITIVE NUMBER.

(+800,806 = 800,806)

ALBERT EINSTEIN STATED THAT IT IS POSSIBLE FOR ONE TO TRAVEL INTO THE FUTURE IF ONE TRAVELS AT THE SPEED OF LIGHT.

## QUESTIONS

1. WOULD YOU LIKE TO TIME TRAVEL?

2. FORWARD OR BACKWARD?

3. HOW MANY YEARS FORWARD OR BACKWARD?

4. WHO OR WHAT DO YOU WANT TO VISIT?

5. WOULD YOU TRAVEL WITH SOMEONE OR BY YOURSELF?